Emotional Creation Theory

Anger - Scared – Happy - Sad

By

Bell Albert

Emotional Creation Theory

Emotional Creation Theory

Anger - Scared – Happy - Sad

By

Bell Albert

Emotional Creation Theory

© 2014 Bell Albert. All rights reserved.
ISBN 978-1-312-11339-8

Emotional Creation Theory

Contents

Dedication

Preface

Chapter One: Evolution: Humans

    Evolution

    Game

Chapter Two: Human Brain: Highlights

    Occipital Lobe

    Hippocampus

    Amygdala

    Insular Cortex

    Frontal Lobe

Chapter Three: Emotions

    Anger

    Scared

    Happy

    Sad

Chapter Four: The Routes

    Emotional Route

    Innate Route

Chapter Five: Facial Nerve Highlights

Chapter Six: Muscles*

    Muscle Progression

    Anger

    Happy

    Scared

    Sad

    Ekman, Paul

Chapter Seven: Emotional Creation Theory

    Mirror Neurons

Chapter Eight: A Minor Tangent

References

Emotional Creation Theory

Dedications

John 3:16

and

Kimberly Lea

"Psychologists, at least psychologists who write text books, not only show no interest in the development and origin of love and affection, but they seem to deny its very existence."

- Harry Harlow

Emotional Creation Theory

Preface

As the reading process shall resume, the reader will note that the book's progression. The starting chapters will inform the audience with necessary conceptualizations and ideologies. However, with each segment, introduction to one's Emotional Creation Theory increases.

Chapter One

Evolution: Humans

Evolution

As Homo sapiens have tried to define various methods, objects, variations and ideations that elude them, one has met evolution. General evolution is the ideation that species have adapted traits that either yield survival (benefit to specimen) or death (did not behoove the specimen); if the specimen were to survive, statistically the beneficiary, the genetics proceed to pass to the processors. However, the needs, environments, and competition are shifting variables. Ergo, advances shall be made specifically to each generation, and the modified genetic changes are passed; thus an evolved species in a scale appearance.

Example: Humans had to adapt to the "fear/ anger" survival skill, for shelter was meniscal, and morality was modicum in existence. Statistically, those of who lived fearlessly were deceased in a brief life span, as opposed to others who adapted a fear conceptualization. The fear acted as caution, a guard that made connections to recognize situational hazards. The survivors passed this profound insight.

Game

Homo sapiens and objects are intriguing, in that they have thoughts, a subconscious, dreams, movement, theories, ideas, interpretations, tangents, or are riddled with numbers, mysteries and an unknown.

Game Theory is the ideology of decision strategy amongst confrontation and cooperation in which the more optimal choice is sought.

Chapter Two

Human Brain: Highlights

Occipital Lobe

The interpreter of the visual stimuli is contrary to the belief that the "eyes see all". The occipital lobe interprets the data gathered from the eyes. It formulates logic, in that it nulls "irrelevant" information, and "zooms" in on caution, wonders and intrigued objects. The brain essentially trains the eyes to perceive threats as "fast" objects, vibrant colours (red, orange, yellow), multi – segmented insects (spiders), unknown, and certain facial expressions (snarl).

Example: An adolescent is tapped on the shoulder by a stranger; the adolescent attends to the tap and looks back. The stranger states, "Boo"; the adolescent is in fear. As the adolescent is trying to rationalize the situation, the brain rapidly computes either a fight, flight, stay, or become familiar computation. The adolescent runs. Ergo, the brain perceived this as a threat and reverted to the adaptation flight to safety and regain basic needs.

Emotional Creation Theory

(Figure 1)

Hippocampus

       The Seahorse Monster

     Hippocampi processing partakes in hard drive actions. Computers store data, as do the brains of subjects. The memory obtained lies in files, vaults, and documents. As Homo sapiens are placed in a situation, all data is grabbed to create logic. The memory (dependent upon rather episodic) tells the Homo sapiens what the situation is, how to interpret it, and what to do. *

Example: An adolescent learns the piano. When he shall play, he forms a subconscious memory of fingering and a conscious thought. (How to play, subconscious; what is being played provides conscious memories of past events).

    * There are eight main sections\ associations of the hippocampus that shall be noted:

CA 1 – CA4

Dentate Gyrus

Entorhinal Cortex

Perhinal Cortex

Emotional Creation Theory

Olfactory Bulb

Amygdala

    Beep, beep; danger, danger; too close

The almond, statistically well known throughout society, partakes in endeavors such as analyzing objects, filtering data, and determining rather they are harmful or anodyne. The amygdala trains the brain lobes to interpret harm, objects, and provides supplementary help to directing data. Along with the occipital lobe, the amygdala refers the data to the hippocampi; as the amygdala recognizes the innate fears, it learns the non – fears, and shall recall them for later references.

Example: An adolescent notes a spider beside a Homo sapiens. The adolescent chooses flight, runs, in order to remain rational. Later, when the adolescent notes the Homo sapiens, (recall, beside the spider) the adolescent became fearful. He must note that either he becomes routed with another fear, or he must recognize that the Homo sapiens was a mere object in the situation and irrelevant. The adolescent became conscious of this decision, and incorporated it into his system; as other spiders occurred, he used a little game to analyze Homo sapiens to yield the following:

Homo sapiens are not the optimal fear. Rationally, they are needed; one must work alongside them to obtain necessities. However, spiders serve no purpose in a cleanly house, for they consist of various poisons and harm (kill) Homo sapiens. Ergo, the recognition of learned and innate fear, (h2o, fire, uneven lines) must not come from choice, rather a process of game. Does the fear benefit or provide *maximum optimization* for the *individual*?

Insular Cortex

"Awe inspiring", how is this feeling described? One does not proclaim to understand the emotional aspects; however, one shall provide insight. The insular cortex, in part, gives others an emotional sensation, an attachment, ("motherly bond") a profound impact of historic remarks, what other went through, overcame and sympathetic and empathetic channeling. Albeit this may seem innate, it is not. The insular cortex is part of the first route variant for emotions. (To be seen in chapter 5)

Frontal Lobe

As the personality holder, and essential emotion maker, the frontal lobe consists of an emotional game process. As noted on page 5, game is a decision process. When a situation is displayed towards a Homo sapiens, the *general* human will provide an emotional and\ or social context process as a decision is needed.

Example: A mother must depart from her 18 years of age child, for he has been accepted to a profound college with a scholarship. The mother disagreed, and began crying. Emotionally, she perceived the loss of him; no more "bear hugs", etc. Then, she became cognizant of her yen, and reverted to "social norms". The mother displayed that socially, society must represent their being in a certain manner, even though the manner may not be the more optimal. Result: The mother had developed an attachment. Her brain went through a different process, an emotional, yet decisional one. As her emotions retreated to the frontal lobe, personal and society order became more relevant. Ergo, she approached the situation with an emotional route, and adjusted socially. With this stated, there are emotional and innate routes. The emotional route is happy and sad, as the innate route is anger (fight) and scared (flight).

Emotional Creation Theory

Chapter Three

Emotions

Anger

By the research of one, *anger* is a psychophysiological response in which the Homo sapiens body has enlisted. The anger (Fight) response is induced by a situational problem.

When Homo sapiens are confronted by situation, the brain relays the data to the hippocampi and amygdala. If recognition of the harm is found, the brain will process an optimal* strategy and proceed. (Figure 1.2 B)

Scared

As an object dependent cause, fear is a psychophysiological response in which the Homo sapiens body has enlisted. (Figure 1.2 B)

Happy

One has theorized that *happy* is a psychological response in which the Homo sapiens mind has induced. As subject 2 is introduced to

---

*In regards to emotions, optimality is an individual manner. Homo sapiens differ in situations due to prior involvement and outcome.

stimulus, a robotic device, he displays a smile. Subject three thinks of church, and begins to smile. Subject four remembered a "riveting" book, and smiles. These subjects are smiling on a conscious level; they are recalling a favorable outcome day, thinking of a beneficial and effective outcome to occur, or experiencing completion. (Figure 1.2)

Sad

Albeit *sad* may appear to be the opposite of *happy*, *happy* is quite relevant to the determination of *sad*. Just as *happy*, *sad* is a psychological response in which the mind induces. These responses produce physiological tangents, as well; however, *happy* and *sad* are psychologically derived, thus classed differently than that of *anger* and *scared*. Ergo, this creates an intriguing system. (Note figure 1.2)

Chapter Four

The Routes

Emotional Route

Receiving input from the paraHippocami, insular cortex, and anterior cingulate gyrus, (Social\ emotional memory) the input goes through the beginning of a route, Route P. (Figure 1.2)

Perhinal cortex, ethorinal 2, dentate gyrus, CA 3, CA1, subiculum (lead to output) **

Hippocampus

Amygdala

Basal Ganglia

Pre - fontal Cortex

Thalamus – Hypothalamus

Orbito - Frontal Cortex

ANS PNS CNS

Pituitary Gland

** As route P surfaces the subiculum, the data is passed, ingrained, and translated in these regions.

Example: A mother developed an attachment, an unconditional "love", and warmth towards her son. As her son retreated to college, she was left without him, only reminiscent objects sustained. When the mother came to clean, she noted a green gravy pan, and began crying. The son had created gravy biscuits the night prior to his departure. She remembered the gravy pan as an emotional object. – Why? As she see noted it, the brain made the connection of cooking, the pan and her son– she remembered the last meal her son created. Ergo, the thought of no son equates to emotional sadness, thus a production of route p to physiological symptoms created by the nervous symptoms.

Emotional Creation Theory

Figure 1.2

Emotional Creation Theory

The Innate Route

The monitor of reactions for Homo sapiens has evolved and anger\fear are innate. The brain has adapted a system in which to remain at homeostasis and survival. By a blueprinted design, the brain seeks to alert an individual of either a learned harm, or a perceived harm.* Rather the situation is constituted to be that of an emotional gathering, as well, the installed route will take priority; thus the individual will act accordingly, then seek an internal emotional evaluation.

Constructed for efficiency, the route remains dependent upon rudimentary functions.

Hippocampus and amygdala (lead to route)

E3, CA1, subiculum (route leads to output)

Madeula Oblongata

Thalamus – Hypothalamus

Neo - Cortex

ANS PNS CNS

Pituitary Gland

*Recall the Occipital Lobe section

Example: An individual wonders a trail; a snake is across from her. The adult displays increased heart rate, sweat, mydrasis, and runs away. Why has she done this? She has recognized fear in the snake, thus a psychophysiological reaction. Her body created sweat to reduce risk of stroke, increased heart rate for oxygen as she shall run, and focused eyes to note attention to the object.

Emotional Creation Theory

Figure 1.2 B

Emotional Creation Theory

Chapter Five

Facial Nerve Highlights

Cranial Nerve 7

As the carrier of axons, the facial nerves partake in facial muscle movement. There are five branches of the facial nerves

Temporal

Zygomatic

Buccal

Marginal Mandibular

Cervical

Emotional Creation Theory

Emotional Creation Theory

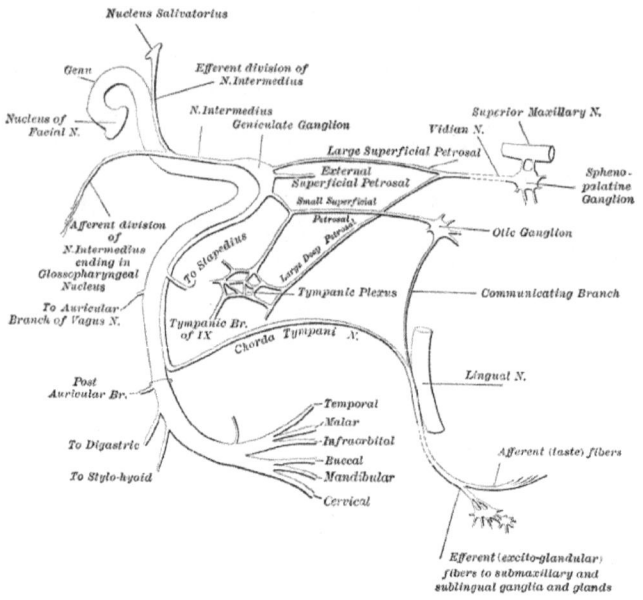

Chapter Six

Muscles *

Anger – Scared - Happy – Sad

Muscles Adaptations

As Homo sapiens adapted more facial muscles, emotional beings were in process of existence.

As the chronological span increases, as do the noted muscles connections. One shall note the intriguing muscles used in innate psychophysiological and learned psychological routes.

Anger

    Orbicular Oris

    Depressor glabellae

    Depressor superculli

    Corrugator suprcilli

    Levator super palpebrae, super tarsal muscle

**FACS: Facial Action Coding System*

Emotional Creation Theory

Happy

    Orbicular Oculi

    Zygomaticus major

Scared

    Frontalis

    Depressor glabellae

    Depressor superculli

    Corrugator superculli

    Risuris (with platsyma)

    Orbicular Oculi

    Masseter

    Temporloris

    Pteryoid

    Levator super palpebrae, super tarsal muscle

Sad

    Frontalis

    Depressor glabellae

    Depressor superculli

Corrugator superculli

Triangularis

Muscle Progression

As seen, one may hypothesize that facial muscles were in mere existent prior to any variant emotion, but when anger (fight) was adapted, a string of responses were to follow. Thus, one may theorize that due to anger, being scared was a logical progression for survival; as the usage of muscles were overtly used, facial nerves began to embed themselves onto neighboring muscles. Ergo, the newly blueprinted and integrated nerves and muscle created emotions.

Emotional Creation Theory

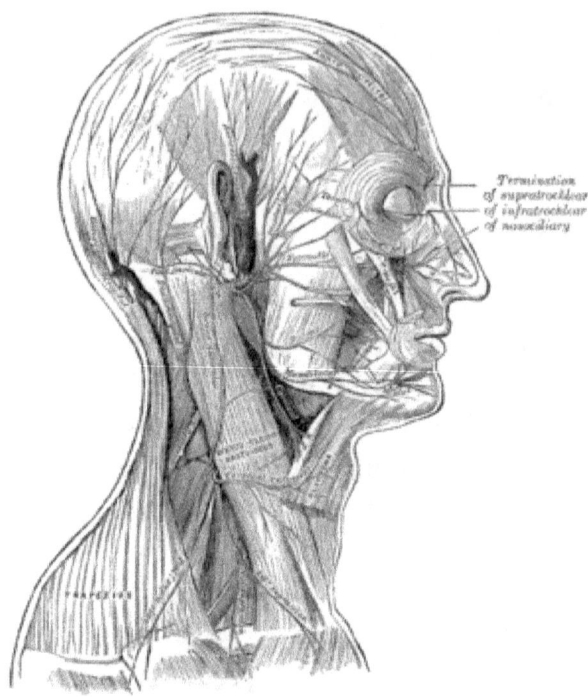

*Reproduction of Gray's Anatomy

Ekman, Paul

Alongside with his fellow mentor, Ekman dove into the unknown of emotional context. As Ekman reached the realms of facial muscles, emotional reactions and new analysis, he notes a tad discovery. He noted that a Papua New Guinea tribe, with no outsource of emotional context seen (no television, frequent travelers, magazines) displayed the similar muscle movement for emotions including anger, scared (fear), happy, and sad. As this was discovered, one elaborated the process; thus the emotional creation theory.

Chapter Seven

Emotional Creation Theory

Mirror Neurons

As infants gain human relations, they develop synaptic connections. Mirror neurons engage the individual to partake in gazing, and mimicry of others. When the infants mimics, muscle formation is gained, as are social abilities. The muscles movement, alongside the memory pertaining as to the reasoning, creates an emotion route.

Example: Day and night, a mother smiles at her infant. As the mother holds the infant, she smiles, and speaks to it. When the infant mimics the mother, there are various underlying functionalities occurring. The infant is associating warmth with smiles; he is forming the basics of emotions, connecting neurons to various sections, forming synapses, and elaborating what such smile means.

If an individual is given a set amount of mirror neurons, and they are not accessed, built upon, then the individual remains emotionless, until the process is enlisted. As the individual proceeds to adulthood, the emotions may still be learned, for there is no age limit.

However, if the emotions are not learnt, then this regards the theory of mind. Emotions allot Homo sapiens to "feel" on some intangible level but not necessarily understand others, rather "connect" to there "level". One does not proclaim this to be accurate for all, for one does not know this "level" or emotions, but one provides observational insight. Ergo, if an individual were to remain emotionless, they would use studies, input, and deduction (essentially a computational theory of mind) in order to try to understand an emotional being.

Emotional Creation Theory

The theory in which a plausible explanation is brought to the reasoning regarding emotions is Emotional Creation Theory. One may gather that emotions were blind adaptations of the Homo sapiens. As Homo sapiens were adapting the usage of anger and scared, the facial muscles were more developed. With the progression of muscle usage, the individuals were creating more in depth facial connections (nerves and muscles). These connections led to a new route. With the emotional route formulated, humans were in need of a way to reproduce it, thus mirror neurons. Mirror neurons complexly create the emotional route connections

Emotional Creation Theory

within the brain and allot an individual to develop emotional ground. Ergo, there is an emotional route, and an innate and more rational route.

A Minor Tangent

Theory of Mind

General Homo sapiens seem to have an understanding of other beings. Is it due to similar situations or is it learned? Are empathy and sympathy the bounds of theory of mind? Psychopathology states that psychopaths lack empathy. Are psychopaths genetically altered, mutated, producing complex cells, hormones, neurotransmitters, or are they psychological produced differently with less routes?

Happy and Sad

Homo sapiens connect warmth with happiness. Logical reasoning may deduce that this is due to the warmth of another individual holding them, warm milk, a blanket, etc. However, what if happy and sad were instilled differently: Happy equates to cold, sad warmth.

By

Bell Albert

References

Theory of Mind

    https://britanica.com/

Gray's Anatomy

    Adapted from figure 790, Henry Gray, gray's anatomy, science

    Adapted from figure 788, Henry Gray, et al., gray's anatomy, science

Prezi

    https://www.prezi.com/

Ekman, Paul

    https://wikipedia.com/

Emotional Creation Theory

Emotional Creation Theory

Emotional Creation Theory

www.ingramcontent.com/pod-product-compliance
Lightning Source LLC
Chambersburg PA
CBHW072302170526
45158CB00003BA/1152